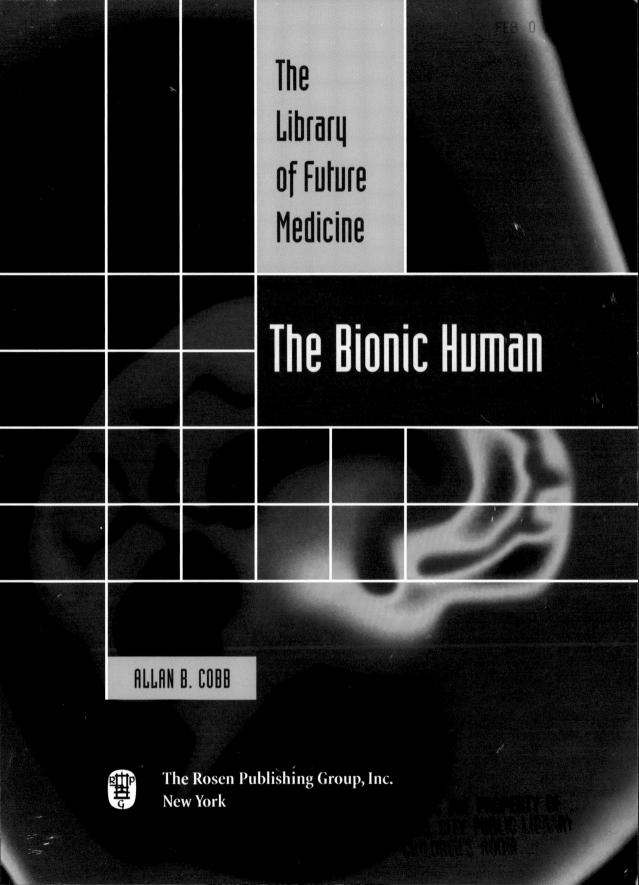

The
Library
of Future
Medicine

The Bionic Human

ALLAN B. COBB

The Rosen Publishing Group, Inc.
New York

Published in 2003 by The Rosen Publishing Group, Inc.
29 East 21st Street, New York, NY 10010

First Edition

Library of Congress Cataloging-in-Publication Data

Cobb, Allan B.
The bionic human / Allan Cobb.—1st ed.
        p. cm. — (The library of future medicine)
Includes bibliographical references and index.
Summary: Examines some of the latest developments in the replacement of damaged human organs and other body parts with artificial or biological materials.
ISBN 0-8239-3670-8
1. Artificial organs—Juvenile literature. 2. Bionics—Juvenile literature. 3. Prosthesis—Juvenile literature. [1. Artificial organs. 2. Bionics. 3. Prosthesis.] I. Title. II. Series.
RD130 .C63 2002
617.9'5—dc21

                                                                        2001006461

Manufactured in the United States of America

Cover image: A man with an artificial arm playing a piano

Contents

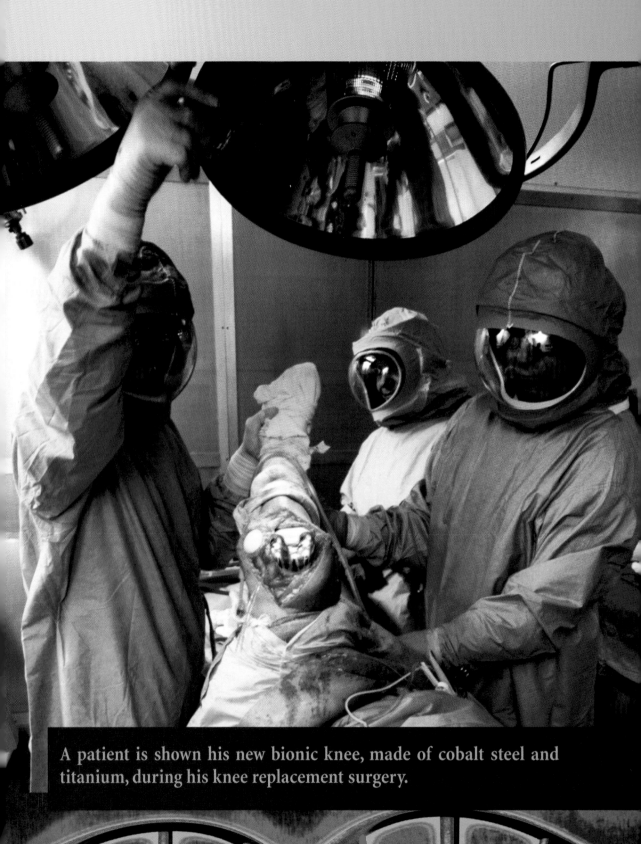

A patient is shown his new bionic knee, made of cobalt steel and titanium, during his knee replacement surgery.

Introduction

In the early 1970s, one of the more popular television shows was called *The Six Million Dollar Man*. The hero of the show, Steve Austin, was an astronaut who had been severely injured in an accident. He was "rebuilt" using bionic technology. He had two bionic legs, a bionic arm, and a bionic eye. His bionic legs allowed him to run at 100 kilometers per hour, his bionic arm gave him incredible strength, and his bionic eye gave him superior vision. This was done at a cost of six million dollars. Steve Austin used his superhuman, bionic abilities as a spy for the government.

Even today, thirty years after that show, the state of medicine is nowhere near as advanced as the rebuilding of Steve Austin. Today, people are fitted with bionic legs that allow them to walk and bionic arms that allow them to perform simple tasks. While none of today's medical accomplishments are as fantastic as those of the 1970s TV series, they are nonetheless remarkable. In the chapters that follow, you will read about some of the amazing things that can be done with bionics and how these remarkable medical breakthroughs came about.

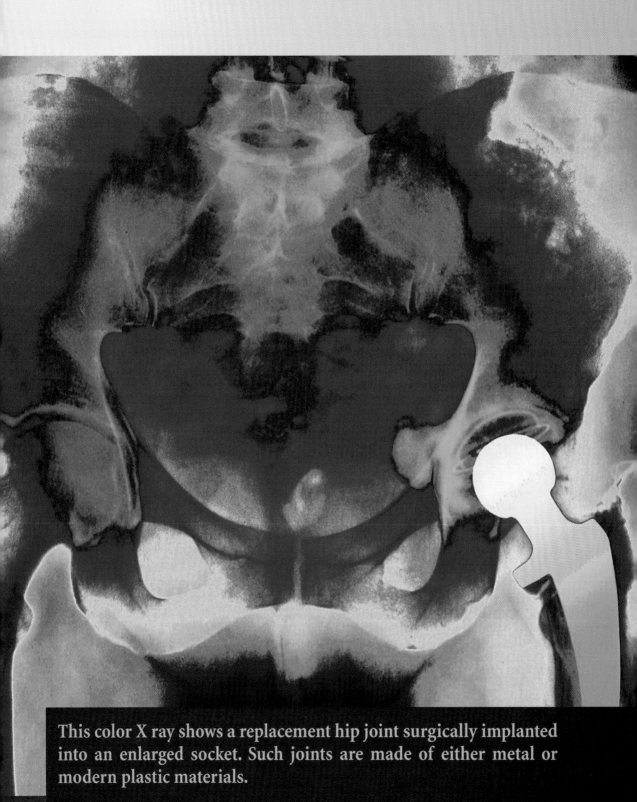

This color X ray shows a replacement hip joint surgically implanted into an enlarged socket. Such joints are made of either metal or modern plastic materials.

What Is Bionic Medicine?

Bionic medicine is a field that studies anatomy and physiology to find ways of imitating or enhancing biological functions with mechanical or electro-mechanical devices. While bionic medicine may sound like science fiction, it really isn't. Bionic medicine is a field of study that has been around for hundreds of years. Bionic devices do not need to be electronic devices; they only need to fix some failure in a biological system. Most likely, you or someone you know could be considered a bionic human. You may know someone with an artificial leg or arm. You may know someone with a hearing aid. Even if you don't know anyone like that, it is likely that you or someone you know wears glasses. These people are all bionic humans.

Any artificial—that is, man-made—part or device that is used on a person is called a prosthesis. A prosthetic leg may be as simple as a wooden peg leg on a pirate or as elaborate as the latest titanium-framed, computer-controlled artificial leg. Some prosthetic devices look absolutely lifelike while others look like part of a robot. Regardless of how a prosthetic device looks, it is its function that is its most important characteristic.

BIOMATERIALS

Biomaterials are any materials used in bionic or prosthetic devices that are biologically compatible with human tissues. For hundreds of years, doctors used a wide variety of materials in their attempts to fix the human body. They tried everything from horsehair to gold wire. Some materials worked while others didn't. In the 1950s, the developing space program caused a boom in the availability of biomaterials. Biomaterials fall within three broad categories: metals, ceramics, and polymers.

Metals are one of the most important of all biomaterials. Metals are strong. They are used for many different applications: as screws to hold bones together while they heal, as the weight-bearing frames of artificial joints, and as strengthening rods, reinforcements, and sutures. Metal hardware may be used for applications that last anywhere from a day to a lifetime. The metals most suitable for use as biomaterials are stainless steel, titanium, platinum, tantalum, and cobalt-chromium alloys.

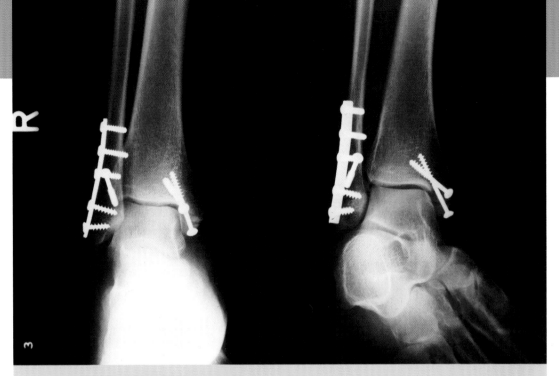

This X ray shows artificial pins being used to hold a patient's ankles together as they heal. Pins such as these are usually made of metal, making them strong and durable.

Ceramics are another commonly used biomaterial. Ceramics are composed of nonmetallic minerals, such as earth, clay, or sand. Ceramics used as biomaterials were first developed in 1969. They make ideal biomaterials because the body does not recognize the ceramic material as foreign and does not reject these parts. Ceramics are used in strengthening bones, replacing or repairing teeth, and making artificial joints.

Polymers are long chains of simpler molecules, called monomers, that join together in repetitive structures.

Cellulose is a polymer that plants produce by creating chains of simpler sugar molecules. But most polymers used in medical applications are petroleum by-products, that is, various forms of plastic. Biopolymers usually belong to one of two classes. Elastopolymers are soft and flexible, and they can be stretched out, but they return to their original shape. Elastopolymers include polyurethanes. Some elastopolymers can break down after prolonged exposure in the body. In some cases this is an advantage, but in others it is a disadvantage. Elastopolymers that break down would be useful in applications where they are needed for temporary support only. The other class of common polymers used as a biomaterial are the true plastics. Plastics can be molded into almost any shape, and they can be chosen for specific characteristics. Some plastics, such as Teflon, are so slick that bacteria and blood clots do not stick to them. Other plastics, such as Dacron polyester, are porous, and tissues can grow into and through them. The uses of plastics as a biomaterial are staggering.

Another important class of polymers are the organosilicon polymers. These are typically referred to as silicone. Silicone is a gel-like or rubbery polymer that is used as a base in facial reconstructions and as a skin on prosthetic devices. The silicone used as skin on prosthetic devices

Artificial eyes and facial prosthetics are made of plastics and silicone, as shown in these examples on casts.

looks and feels much like real skin. This is valuable for giving artificial arms and legs a natural look.

New biomaterials are being developed all the time. The latest biomaterials are composites. Composites are made up of materials from two or more of the above groups. This may include metals that are covered with plastic or plastic that is reinforced with carbon fibers.

PROBLEMS WITH BIOMATERIALS

Material selection is extremely important when deciding which biomaterials to use for a medical application. For

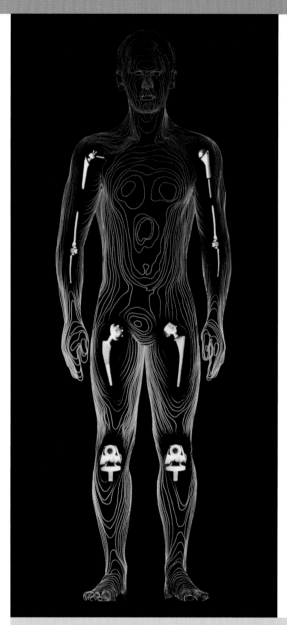

example, if two metals are used together in the same application, they must be chemically stable. Some metals, if they are touching or very close, can cause each other to corrode. The residue from this corrosion may be toxic, or it may cause problems with the functioning of the device or cause it to fail altogether. If plastics reinforced with carbon fibers wear out, the carbon fibers released could cause cancer. The materials that are chosen for prosthetic devices are as important as the design of the device itself.

This idealized medical image shows the typical locations for titanium bone and joint replacements.

REPLACEMENT JOINTS

Replacing joints has become a fairly common medical procedure. Replacement joints may be needed because of physical injuries or conditions such as advanced arthritis. Replacement joints are made from a variety of biomaterials, depending on the application. High-impact joints, such as a hip joint, are made from metal for durability. Other joints, such as a wrist joint, are made from plastic or plastic-coated metal to reduce friction so that the articulation of the joint will be normal.

Replacement joints are usually designed to look and function identically to the originals. For example, the hip joint is a ball-and-socket joint. A replacement hip joint looks exactly like the original. The original joint is removed from the bone, and the replacement joint is attached to the bone and then put back into place.

THE SKELETON

The human skeleton protects the soft internal organs, provides attachment points for the muscles so that we can move, and provides a strong framework so that we can stand upright. Bones will usually heal if they are broken. It is not uncommon to see people with casts on their arms or legs.

Holding the bone rigid allows it to redeposit calcium and return to its original strength and shape.

The skeleton is held together by ligaments and tendons. Ligaments hold bones together at joints and allow flexibility, while tendons attach muscles to bones. Ligaments can be damaged as a result of accidents, but can be repaired or even replaced with biomaterials with varying degrees of success. Ligaments are usually repaired with plastic mesh or yarn that can be stretched. This allows healthy tissue to grow on the biomaterial so that it functions almost naturally.

Tendons can also be torn as a result of accidents. Tendons are more easily repaired than ligaments because tendons do not need to stretch like ligaments. Tendons are usually

An artificial knee ligament made of Gore-Tex, a new synthetic fabric

repaired with low-stretch plastics. The strength in tendons is needed because muscles are what give us the ability to move.

MOVING

Movement results from muscles that contract. The muscles responsible for a particular movement are always paired with muscles that control the opposite movement. Muscles are attached to bones with tendons, and when they contract they pull on a bone. When muscles relax, they simply cease to pull on a bone, rather than move it back to its original position. Another muscle is required for that. The muscles that move the arm at the elbow are the biceps and the triceps. When the arm is extended straight out, the biceps contracts and flexes, or pulls, the forearm upward and close to the upper arm. When the arm is flexed like that, the contraction of the triceps muscle pulls the arm straight again.

This type of movement based on contracting and relaxing muscles takes place at every joint. These are important principles to understand because artificial arms and legs must operate in the same way if they are to function properly.

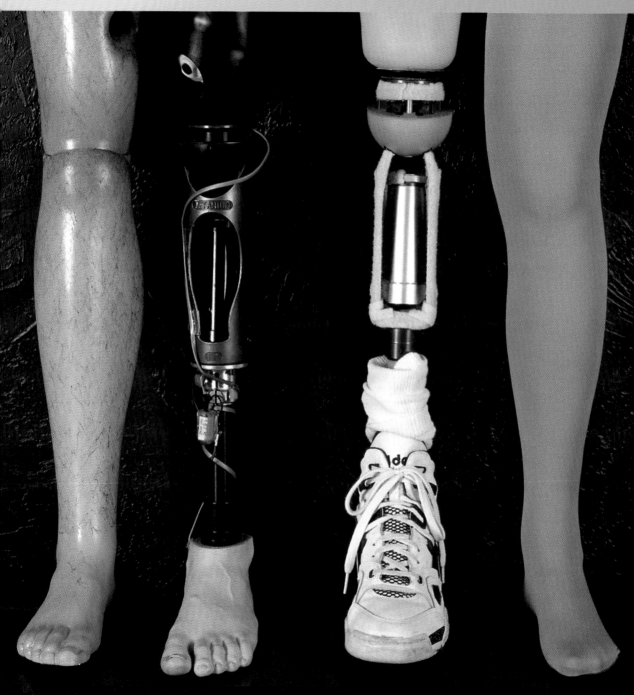

Prosthetic legs at the Sabolich Prosthetic and Research Center in Oklahoma City, Oklahoma. Each uses different materials and coverings.

Bionic Limbs

The loss of a limb usually results from some type of accident. If a limb is badly damaged, it may require amputation to prevent serious infection. Some diseases or conditions may also require the amputation of a limb. Sometimes, a limb may be missing as a result of a congenital birth defect. There are thousands of people who could benefit from well-designed bionic limbs.

LEGS

Prosthetic legs have been around for hundreds of years. The earliest prosthetic devices were quite simple and crude, like the peg leg, but they did serve the purpose of restoring the ability to walk. These crude prosthetic devices were most useful to patients who lost a leg below the knee. By still having the knee to help move the prosthesis, learning to walk was much easier. If the patient lost a leg above the knee, walking was still possible but more difficult and awkward. All movement of the leg resulted from moving the hip. If the entire leg was lost, a prosthetic device was not especially useful.

CONVENTIONAL PROSTHETIC LEGS

Today, many prosthetic legs for below-the-knee amputees are still made of wood. Wood is lightweight and strong. The wooden legs in use today are much different from the peg legs of days past. Modern wooden legs are shaped like normal legs, complete with feet. The foot has a cushioned heal and flexible toes. These two features allow the wearer of the artificial leg to walk with a more normal gait or stride.

Above-the-knee amputees usually have a more complicated prosthetic leg. Modern prosthetic legs usually have some kind of a joint at the knee. Some have a hydraulic knee that operates like the power brakes on a car. Pressure increases as more body weight is placed on the leg, and a fluid-filled cylinder responds by increasing pressure to stiffen the knee. Pressure decreases as body weight is shifted to the other leg, and the hydraulic pressure decreases, allowing the knee to bend. This provides a steady, fairly normal gait.

BIONIC LEGS

Bionic legs provide a more natural gait by controlling the movement of the leg during activities such as walking. Modern bionic legs utilize computer technology to control how the leg responds and to provide a smooth, natural gait. The computer technology found in one such bionic leg is the same technology that is used by the military to improve stability and performance in aircraft. This bionic leg, called the 3C100 C-Leg System, is built by Otto Bock Orthopedic Industry.

The muscles that are used to control movement of a real leg are replaced with hydraulic units. The hydraulic units are controlled by a built-in microprocessor. The microprocessor is programmed on the basis of a scientific analysis of walking movements as well as biomechanical studies. Sensors in the knee and shin area continuously monitor movement and relay this information to the microprocessor. This feedback system is essential for the fine-tuning of leg movements and mimics the message transmission system of the nerves. Because this data gathering and information processing is instantaneous, the movement of the leg is precisely controlled. The stability that the bionic leg provides is a result of the way the knee joint is controlled. When the foot touches the ground, the hydraulic units stabilize the knee and prevent additional rapid

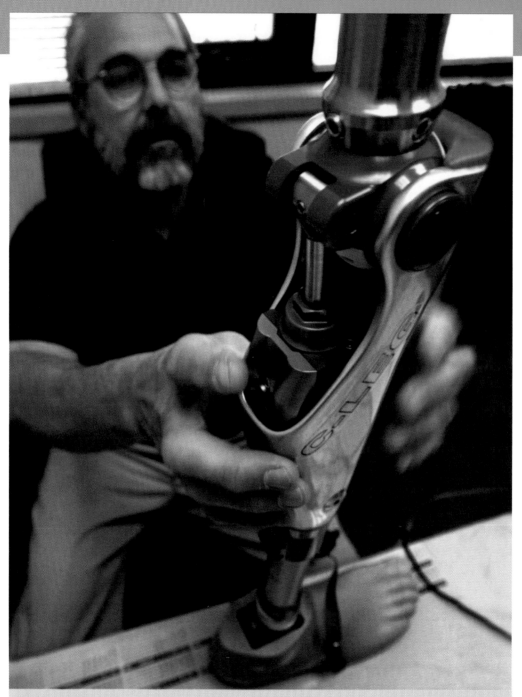

A prosthetist examines a 3C100 C-Leg System. The C-Leg has a computer microprocessor that helps control the movement of the artificial limb when it is attached to a person.

movement. This means that the person wearing the leg can easily walk up or down stairs or inclines and walk at varying speeds with ease even over uneven terrain. The microprocessor also has the ability to be custom programmed for the user.

The Bock bionic leg is constructed of lightweight carbon fiber material. In fact, the entire bionic leg weighs only 1.1 kilograms, a little more than 2 pounds. A rechargeable lithium-ion battery powers the microprocessor and the hydraulic units. The battery will power the leg for twenty-five to thirty hours on a single charge.

HANDS AND ARMS

In the past, an amputated hand, or even a missing lower arm, was usually replaced with a crude hook or some other such implement. While the hook did not have any movement capabilities, it did allow the user more freedom than having nothing at all. This type of prosthetic device was most useful if the loss occurred below the elbow. Having a functioning elbow gave the user a reasonable, although limited, range of movement.

CONVENTIONAL PROSTHETIC ARMS

Prosthetic arms and hands are much more complicated than prosthetic legs. The prosthetic leg only needs to move forward and backward. Lateral, or side-to-side, movement is not as critical. The arm, however, is required to move forward and backward and out to the side. If the shoulder is intact, these movements are easier. The hand is used for grasping and manipulating objects, and it must also be able to rotate at the wrist. This rotational ability is necessary even for simple tasks like turning a doorknob.

The first totally electric prosthetic arm, called the Utah Arm, was made available in 1981. The arm featured a totally electric elbow and hand. It was made of injection-molded plastic with nylon and carbon fiber reinforcements. In all, the arm weighed approximately 1.5 kilograms. Electrodes in the arm were designed to read myoelectric signals from the muscles in the upper arm. A microcomputer interpreted the signals

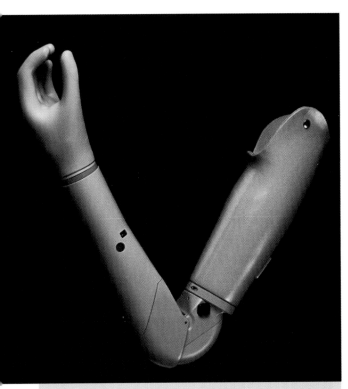

The Utah Arm was developed by a team led by Dr. Stephen Jacobsen at the University of Utah's Center for Engineering Design.

and translated them into arm movements at the elbow. The movements were controlled by tiny electric motors. They were powered by an internal, rechargeable battery pack.

The arm had several disadvantages. It could not be lifted very high or moved close to the body when raised. The arm could not be used for a simple action like drinking from a cup of water because it lacked the ability to rotate. The arm did have advantages over non-powered prosthetics in that its movements were subtle and lifelike.

The hand used on the arm was called the Otto Bock Hand. The Otto Bock Hand was controlled by a myoelectric system that responded to electrical signals from muscles higher up in the shoulder. The motions of the hand were limited to grasping. Batteries stored in the forearm powered the hand. An on/off switch was located in the palm. The hand was modular and could be quickly removed and exchanged for a variety of different tools.

The design and operation of prosthetic arms has remained relatively unchanged since the Utah Arm and the Otto Bock Hand. The design has been refined through the addition of lighter, more powerful motors, better sensors, better microcomputer controls, and lighter, more powerful batteries. The coverings for the arm have changed from plastic to a more realistic looking silicon. However, many of the problems associated with its limited range of movement have not been solved.

Fitting children with prosthetics has been made easier by the development of lighter materials.

BIONIC ARMS AND HANDS

The development of the bionic arm has actually been a gradual refinement of an old design. The addition of new materials and better technology has produced an ever-improving bionic arm. Doctors are now able to attach the prosthetic arm at the shoulder blade, and they have improved the myoelectric contacts, but the basic design is still the same. The real advance in the bionic arm has been with the bionic hand.

The hand is an incredible tool. About half the bones in the human body are found in the hands and feet. The fingers have an incredible range of fine, delicate movement. The electric-powered hand was developed more than fifty years ago. It was designed simply to grasp. The new electronic hand with micro-processor controls has greatly miniaturized components within the individual fingers, so that very delicate movements have become possible. Natural wrist movements have been successfully duplicated. Smaller, lighter, more efficient motors, hydraulics, and batteries brought about further improvements.

POWERING BIONIC DEVICES

Bionic devices need electrical power to run the microproces-sors, motors, and hydraulic devices within them. Batteries usually supply this power. The earliest batteries used on bionic limbs were heavy and had a short lifetime. This meant that they needed to be recharged frequently. Today, battery technology is much more advanced. Lithium-ion batteries have a high energy-to-weight ratio and recharge quickly. Bionic arms and legs are large enough so that they have plenty of storage room for the batteries. Other bionic devices, such as bionic eyes and ears, are smaller, and they have to overcome the problem of miniaturizing the power supply.

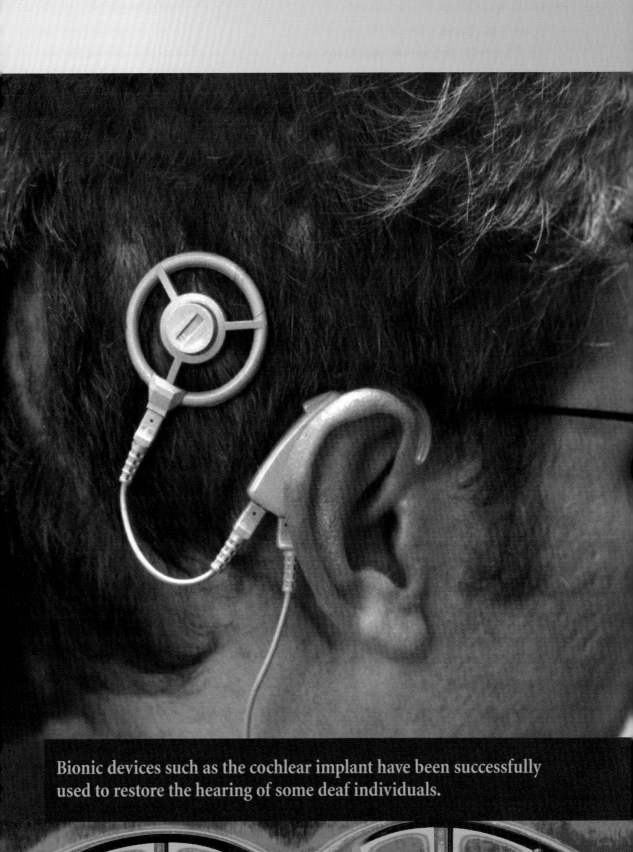

Bionic devices such as the cochlear implant have been successfully used to restore the hearing of some deaf individuals.

3

Bionic Eyes and Ears

Vision and hearing are two senses that most people take for granted. However, millions of people suffer from diseases or conditions that leave them blind or deaf. In some of these cases, sight and hearing may be restored through bionic devices.

HOW THE HUMAN EYE WORKS

Vision is the result of the stimulation of the nerve cells in the eye. Light enters the eye by passing through the cornea and aqueous humor and then through the pupil. The pupil changes its size depending on the amount of light entering the eye. Immediately behind the pupil is a lens. The lens in the eye is not rigid like the lens of a magnifying glass. Instead, the lens is soft, and its shape is changed by the contraction or relaxation of the surrounding muscles. Like the lens of a magnifying glass, the lens focuses light on the retina.

When the light strikes the retina, it stimulates specialized cells. The retina is made up of two types of light-sensitive cells. These cells are called rods and cones. Rods are light receptors that help you identify movement and shapes. The cones are light receptors that help you detect color. At the back of the eye, all the nerves come together to form the optic nerve. The optic nerve runs to the thalamus in the brain and then to the back of the cerebral cortex, where the images are interpreted by the brain as an image.

BLINDNESS

Worldwide, about 42 million people suffer from blindness. In the United States, about 1.1 million people are blind. Only about 3 percent of all cases of blindness are the result of accidents. The rest of the cases are the result of eye diseases and various conditions that cause damage to the clear portions of the eye, the nerves within the eye, the optic nerve, or the vision centers in the brain.

The most common causes of blindness are conditions that affect the clear portion of the eyes, such as the development of cataracts or corneal clouding. A growth of tissue covers the cornea or the lens, which becomes opaque and prevents light from passing into the eye. Surgery restores vision to patients in about 90 percent of cases.

Another leading cause of blindness is damage to the nerves within the eye. The most common conditions that cause damage to the eyes are diabetes, macular degeneration, and retinitis pigmentosa. Loss of vision from diabetes or macular degeneration occurs because of a reduced blood supply to the central portion of the eye. The blood vessels that supply blood to the rods and cones in the retina become damaged and no longer function. Laser surgery can sometimes slow down the onset of this type of vision loss, but it is not reparable. Retinitis pigmentosa is a degenerative disease that starts in early childhood and is characterized by a slow loss of vision. This disease affects the rods, the optic nerve, and the pigments in the retina. This condition is untreatable.

Damage to the optic nerve or the vision centers in the brain usually results from some kind of accident or trauma to the head. The optic nerve may also suffer damage from diseases, infections, or cancerous growths. Damage to the optic nerve and the brain are usually untreatable once the damage has occurred.

Blindness resulting from conditions that affect the clear part of the eye is easily treated with conventional medical technology. Conditions resulting from damage to the optic nerve or vision centers of the brain are well beyond any current technology. Current bionic research is focusing on blindness caused by damage to the nerves within the eye.

STEPS TOWARD THE BIONIC EYE

The current work on bionic eyes targets people who suffer from macular degeneration and retinitis pigmentosa. Both of these conditions are caused by the failure of the rods and cones in the retina to function properly. Current technology is aimed at circumventing the inoperative rods and cones and stimulating the nerves that are connected to them directly. Researchers are working on two different approaches.

The Retinal Implant Project, a collaborative project between Harvard University and the Massachusetts Institute of Technology, began its research in 1988. The group worked on a retinal prosthesis that rests on the inside surface of the retina near the optic nerve. This retinal implant consists of two microchips. The first microchip is a tiny solar cell that receives a signal and powers the second microchip. The second microchip is the stimulator. It decodes the information received by the first chip and stimulates the appropriate nerves in the retina.

This system requires a special pair of glasses. The glasses have a small electronic camera much like a video camera. The images recorded by the camera are converted

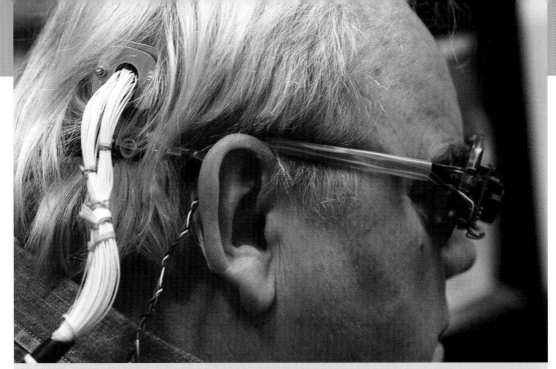

With the help of electrodes implanted in his brain, a camera mounted onto a pair of glasses, and a ten-pound computer, this blind man can read large letters and move around big objects in a room.

into digital signals by a microchip located in the frame of the glasses. A small laser transmits the digital information to the first microchip. The light from the laser creates an electric current in the first microchip that allows the second microchip to decode the digital information carried in the laser beam.

As yet, this system has not been implanted into a human eye. Experiments are still being conducted to find materials that are compatible with the retina, and doctors are still trying to figure out exactly how to connect the device to the nerves in the retina. If this technique is successful, it could allow many blind people to see again.

The second approach to developing a bionic eye is being undertaken by the Optobionics Corporation. Dr. Alan Chow and his brother, Vincent Chow, a mechanical engineer, have developed a self-contained retinal implant called the Artificial Silicon Retina. The implant is 2 millimeters in diameter and 0.03 millimeters thick, and it contains 3,500 microscopic solar cells called microphotodiodes. Each of the microphotodiodes has electrodes that stimulate with an electrical impulse the cells in the retina that still function. Because light falling on the microchip powers the device, there is no need for any external power source, batteries, or wires.

This technology has been tested on animals, and the device does produce stimuli at both the retinal surface and in the brain, though animals cannot tell us how their vision is affected. Testing on humans began in June 2000. The Artificial Silicon Retina was implanted in three individuals. A second human trial was started with three additional patients in August 2001. The results of the clinical tests are still being evaluated and have not been released to the public.

THE EAR

Hearing works because sound waves enter the outer part of the ear and travel down the ear canal to a flat membrane called the eardrum. The energy of the sound waves is transferred to the eardrum, which begins to vibrate. The vibrations from the eardrum are picked up and transferred through three small bones—the malleus, the incus, and the stapes. The vibrations in the stapes are picked up by the fluid-filled, snail-shaped cochlea. The inner surface of the cochlea is lined with hair cells. As vibrations pass the hair cells, the hairs bend. This bending of the hair cells triggers nerve impulses that pass through the cranial nerve to the brain and are interpreted as sound.

Higher frequency vibrations travel further through the spiral passage of the cochlea and stimulate hair cells near its apex. Lower frequency sounds do not travel as far, so they stimulate hair cells closer to the cochlea's base. The brain interprets the frequency of the sound by which hair cells are stimulated.

HEARING LOSS AND DEAFNESS

Partial deafness is the most common type of hearing loss. This is a loss of the ability to hear certain sounds, often the higher and lower tones. This may be a temporary condition or permanent depending on the cause. Hearing loss usually begins gradually and becomes worse as time passes. Total deafness, or the inability to hear any sounds, usually only results from severe damage to the cranial nerve or the brain, or it is a result of a genetic condition.

Hearing loss can be caused by overexposure to loud sounds, by certain medicines, by an infection, or by tumors. When something happens that damages hair cells in the cochlea, hearing loss results. Damage to these hair cells is irreversible because nerve cells do not regenerate. Once they are damaged and nonfunctioning, they will never function again.

For people with partial hearing loss, hearing aids are used to restore some hearing function. Hearing aids are not really bionic devices. They are simply small electronic devices that amplify incoming sounds to make them louder and clearer. Hearing aids do not completely restore hearing loss, but they do make it easier to hear different sounds.

COCHLEAR IMPLANTS: THE BIONIC EAR

For people with total or nearly total hearing loss, cochlear implants are the only technology available that can make them hear again. Cochlear implants were first used in the late 1980s. The cochlear implant does not totally restore hearing, but it does restore a portion of hearing. It consists of a thin bundle of from four to eight electrodes that are surgically inserted into the snail-shaped cochlea.

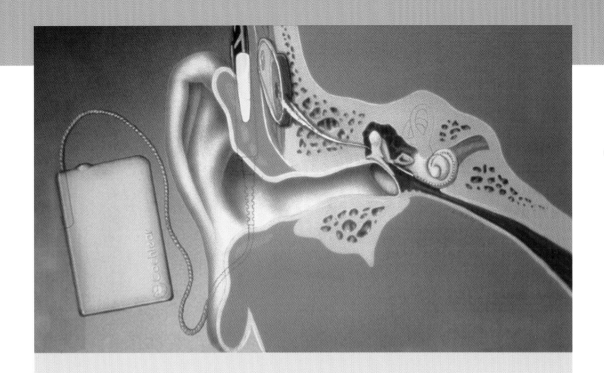

This illustration shows a cochlear implant inside an ear.

A microphone is worn behind the ear on the surface of the skin. The microphone is connected to a sound processor that converts the sound into four to eight different electrical stimuli. The electrical signals bypass the nonfunctioning parts of the middle ear and pass through the skull directly to the inner ear. Because each electrode ends at a certain spot in the cochlea, the brain interprets the electric stimulus as a sound at a certain frequency.

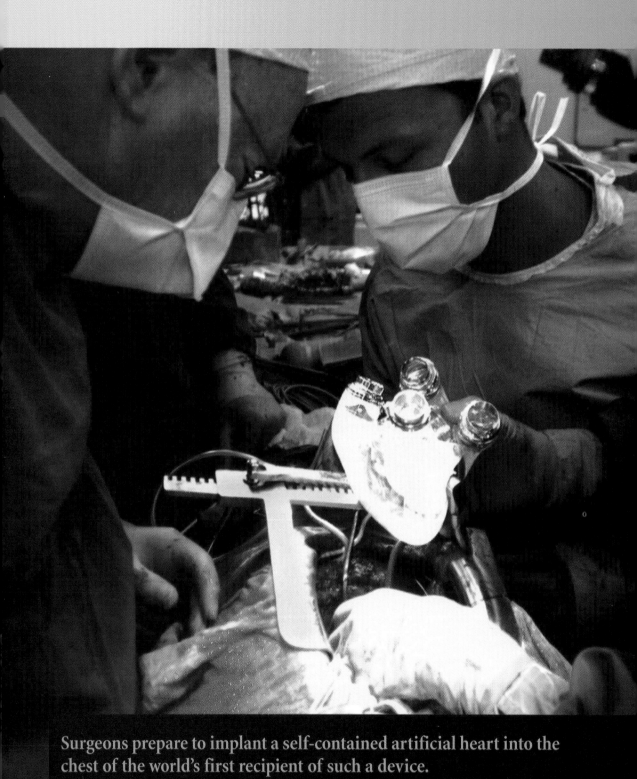

Surgeons prepare to implant a self-contained artificial heart into the chest of the world's first recipient of such a device.

4

Bionic Organs

Bionic limbs, eyes, and ears certainly improve the quality of a person's life. They restore lost sensory or motor functions, but they are not essential to life. Bionic organs, on the other hand, may actually save or prolong life. Organ transplants are becoming common operations. However, donor organs that are compatible with the patient's body are not always available when they are needed, and so the search for satisfactory artificial replacements continues.

Bionic or artificial organs come in a variety of types, depending upon whether or not parts of existing organs are used or the organs are totally replaced. A hybrid organ is an organ that combines natural tissues with artificial materials. Hybrid organs are useful when scientists are not able to duplicate the natural organ's functions. For example, liver cells may be contained in an artificial skin composed of some kind of biomaterial that replaces all or part of a patient's liver. The liver cells carry on the filtering functions of the natural liver, but they are enclosed in a nonliving container. Hybrid organs combine the advantages of artificial organs and organ transplants. Hybrid organs usually have a lower rate of rejection by the patient's body.

Another type of artificial organ is the biosorbable organ. Biosorbable organs are made from artificial materials that act as a framework for natural tissue to grow on. The biosorbable structure is seeded with living cells that regenerate natural tissue. Damaged blood vessels are one application for biosorbable organs. The damaged segment of a blood vessel is replaced with a biosorbable material. As the implanted cells grow around the biosorbable material, they replace its function and regenerate a new blood vessel. The difficulty of using biosorbable organs lies in finding materials that are suitable for the application. The materials must be strong enough for the task. They must not damage the implanted cells or stimulate an immune reaction.

HYBRID LIVERS

The artificial or hybrid liver is an excellent organ to use as an example of artificial organ technology. The human liver is very large. Over half the liver can be surgically removed to deal with disease or damaged tissue, and the liver will, over time, regenerate itself. However, while it is regenerating, it must have some replacement system to carry out its functions. The liver has three general functions. First, it is involved with the regulation of carbohydrates, lipids (fats), proteins, and blood clotting factors. Second, the liver is also responsible for synthesizing a number of different enzymes that are used throughout the body. Finally, the liver filters and removes toxins from the blood. Designing an artificial organ that performs all these functions is a daunting task.

The best approach for designing a functional artificial liver seems to be the development of a hybrid liver. Getting natural liver cells to perform the desired functions within an artificial organ is not an easy task. Liver cells are extremely difficult to culture and maintain in the laboratory. They quickly lose their ability to perform their specialized tasks when they are removed from the liver. Scientists have made some advances in designing a hybrid liver to work with living cells, but its ability to function over the long term has been limited.

HYBRID HEARTS

In contrast to the artificial liver, the artificial heart is a simple device. The function of the heart is to pump unoxygenated blood to the lungs and then pump oxygenated blood from the lungs to the rest of the body. The human heart does this with four chambers and a series of one-way valves. In its most simple form, the heart can be thought of as two pumps, each with two chambers.

Blood from the body enters the heart through the right atrium. Blood from the right atrium drains into the right ventricle through a one-way valve. When the right ventricle fills, it compresses. The one-way valve from the right atrium closes, and blood is forced through another one-way valve toward the lungs. As the blood passes through the lungs, dissolved carbon dioxide leaves the blood, and oxygen is absorbed. The blood then returns to the heart and enters the left atrium. The blood drains through a one-way valve into the left ventricle. When the left ventricle is filled, it compresses in unison with the right ventricle, and blood is pumped through a one-way valve into the aorta and off to the rest of the body.

Each time the heart beats, the right and left ventricles contract and move blood. The heart beats about forty million times per year, and it pumps about five million liters of blood. The rate at which the heart beats varies. When resting or sleeping, the heart beats slowly. During times of strenuous exercise, the heart beats rapidly and pumps much more blood.

Hearts can develop a number of different problems. One of the most common problems is the failure of one of the one-way valves. These valves can be damaged as a result of infection, illness, or a congenital birth defect. When a valve starts leaking, the efficiency of the pumping action of the heart is greatly reduced. This causes the heart to work harder and harder to overcome the inefficiency. This adds considerable stress to the heart. Scientists and doctors have designed replacement valves for the heart that mimic the function of the original valves.

The first heart valve replacement surgery took place in 1952. Since that time, hundreds of different valves have been designed and used. These different designs fall into two main categories: biological valves and artificial valves. The biological valves are taken from pig hearts and attached to plastic rings. The plastic ring aids in the insertion of the valve into the recipient's heart. The valve works identically to a human heart valve. Small overlapping flaps open under pressure and then close off to prevent flow in the opposite direction. These valves are seldom used today because they wear out in about seven to ten years.

There are a variety of different types of mechanical valves made with different materials. The earliest mechanical valves were of a ball-and-cage design. A ball of plastic is retained in a cage. As the heart pumps, the ball is pushed forward in the cage and blood flows around it. As pressure decreases, the ball is pulled back and closes off the valve. The design is very efficient. The drawback is that some blood cells are crushed and destroyed as the valve closes. This promotes unwanted blood clots. Blood clots cause serious medical problems, such as strokes and heart attacks. To combat this, patients with this type of valve are required to take blood thinners to prevent clots.

Today, mechanical heart valves are much more common. The mechanical valves used today are usually valves with flaps much like a natural valve. The valves are made of plastic or titanium. These mechanical valves still damage blood cells and require the patient to take blood thinners to prevent clotting. However, mechanical valves last considerably longer than biological valves.

An artificial heart valve

ARTIFICIAL HEARTS

New research in heart valve technology has led to the development of the complete artificial heart. The idea of a complete replacement heart is not new. The first artificial heart was implanted in a dog in 1957. The heart kept the dog alive for ninety minutes. The development of an artificial heart for humans took many years. The earliest designs required extensive external controls and power supplies. As computer technology developed, the size of the external support equipment was reduced.

The first really practical artificial heart was developed in 1977. This artificial heart was called the Jarvik-7. The Jarvik-7 heart was made of polyurethane reinforced with a Dacron mesh. While not shaped like a natural heart, the Jarvik-7 heart had four chambers. The Jarvik-7 pushes compressed air against a diaphragm in its artificial ventricle. The expanding diaphragm creates the pressure to close the valve from the atrium and force blood out of the ventricle. An external unit controlled the heart rate. The original external control unit weighed almost 150 kilograms, more than 300 pounds. In time the weight of the control unit was reduced to about 5 kilograms, or about 10 pounds. The life span of the Jarvik-7 heart was about 200 million cycles, or four to five years. The Jarvik-7 heart was an important step in the development of artificial hearts.

The natural heart works by contracting its walls to create a pumping action. Scientists do not yet know how to make a container that compresses itself. The current technology for pumps is useless for artificial hearts because the pressures these pump designs create would damage red blood cells. The current technique used to overcome these problems in the latest artificial hearts is to use a flexible bladder inside a rigid container. These new artificial hearts have four chambers like a natural heart. The blood in the ventricles is pumped by a small motor that simultaneously compresses two flexible bladders in the right and left ventricles.

The newest version of the artificial heart uses a control unit that is about the size of a deck of cards. This control unit contains both a microprocessor and a battery. The control unit

Stored in a tank of water to keep it sterile, the Arrow LionHeart heart implant device can function in place of damaged ventricles. The Arrow LionHeart is not an artificial heart. It is a "heart helper" or left ventricular assist system for patients with severe heart failure.

is small enough to be surgically implanted in the abdomen of the patient. The heart is capable of pumping about ten liters of blood per minute. This is enough blood for most activities, except for strenuous exercise. It is hoped that the newest generation of artificial hearts will be capable of supporting the patient for many years. At present, these artificial hearts are used to maintain patients only until a donor heart can be found.

Research into bionic organs does not stop with the heart and the liver. Other organs, such as the pancreas, the kidneys, and the lungs, are also being developed. The development of such artificial organs is difficult because the body's organs each perform a complex series of functions. In some cases we still don't fully understand all the functions of organs, or how various organs interact with each other. There are still issues of immune system reactions and organ rejection, though much progress has been made with new immunosuppressive drugs. Miniaturization of replacement organs is another problem.

It is also possible that new developments in the field of embryonic stem cell research may lead to the growing of replacement tissues for heart muscles, nerves, and other systems of the body, eliminating the need for complex biomechanical devices.

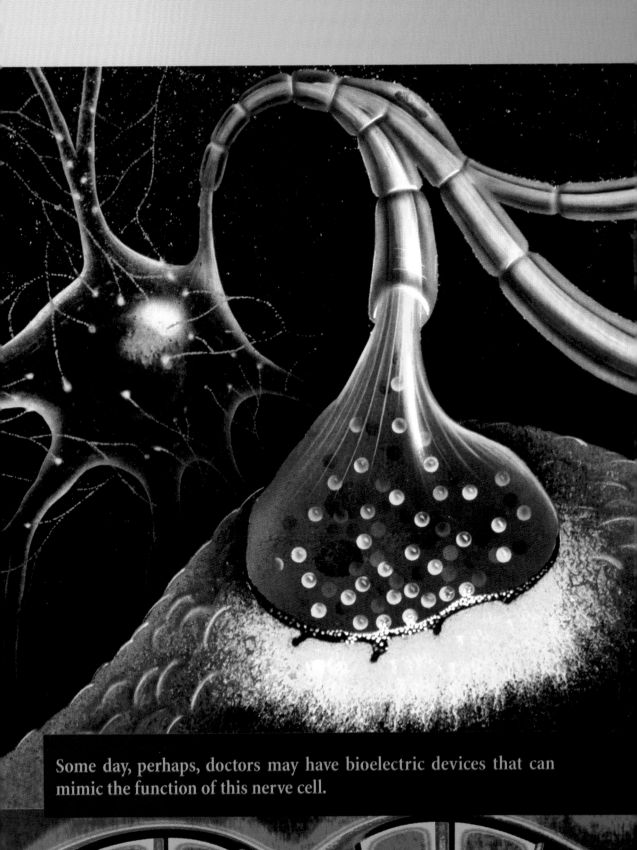

Some day, perhaps, doctors may have bioelectric devices that can mimic the function of this nerve cell.

5 Neural Connections

Imagine being able to turn lights on or off, dial a phone, or type a document at a computer using only your thoughts. These ideas have appeared in science fiction stories for at least fifty years. Today, some of these types of activities are becoming reality.

In 1849, Emil Heinrich Du Bois-Reymond, a German physiologist, detected minute electrical discharges created by the contracting of muscles. This technology was first used in prosthetic arms in the 1970s. Myoelectric sensors were used in artificial arms to control movement. The sensors would "read" the myoelectric signals in the muscles of the upper arm and translate those signals into commands to move the arm. This technique gave profoundly handicapped individuals the ability to move using artificial limbs.

The use of myoelectric sensors in the 1970s was crude. The signals were received and translated into movement but the degree of movement was difficult to control. Sophisticated microprocessors can now "read" the myoelectric signals and interpret the intensity of the signals. This leads to a much more precise movement of the artificial arm. The more precise the movement, the better an individual is able to perform delicate manipulations with his or her arms and hands. The interpretation of these myoelectric signals served as the basis for the next advance in neural connections.

Connections to the muscles to read myoelectric signals are usually accomplished with electrodes that are placed on the skin. The electrodes record the faint signals and send them to a microprocessor that amplifies them, converts the information to a digital format, analyzes the signal, and then sends another signal to a specific motor or hydraulic unit to produce the appropriate motion. The myoelectric signal flows in only one direction, from the muscle to the microprocessor. There is no feedback mechanism to return signals from the microprocessor to the muscle, as there would be with a real system of nerves.

Attempts have been made to connect sensors directly to nerves. Scientists try to attach sensors to nerve cells by inserting thin needles into them. These attempts have been largely unsuccessful in the long run because the needles tend to cause infections or other problems. In addition, isolating individual nerves is difficult as they are often bundled together. The possibility of infecting or damaging the entire nerve bundle is too great.

Scientists are working on new technologies to connect nerve cells to sensors. One of the latest techniques is to combine a nerve cell and a single circuit on a silicon chip. This technology may make it easier to connect a computer to a neural network, and it may lead to computers that utilize the power of nerves. But this is still really in the realm of science fiction. No one can yet say in what way an electrical signal passing between two nerve cells represents an actual thought.

THE EYE AGAIN

Another approach to using the electrical signals of the nervous system is by exploiting the electrical activity that takes place in

Dr. Douglas Kondziolka *(right)* explains the type of device that he and Dr. Leland Albright *(left)* surgically implanted into a juvenile cerebral palsy patient to subdue violent tremors in the patient's arms.

the muscles of the eye. Scientists have designed electrodes that measure the change in electric potential that occurs when the eye changes its position. With a computer, these signals can determine where a person's gaze is directed. This technology has been incorporated into some military weapons. For example, the machine guns on some military helicopters can be aimed by tracking the movements of the pilot's eye.

This is not really the kind of "mind over matter" control we talked about at the beginning of the chapter. The sensors used read changes in the electric potential or voltage of the eye muscles as they contract. To achieve true mind control, a mechanical device needs to be able to interpret signals directly from the brain.

THE MIND-BODY INTERFACE

Since the electrical signals generated by the brain were first discovered, scientists have tried to equate particular behaviors with certain patterns of electrical activity. These electrical patterns are only partially understood because of their highly complex nature. And the mind-body interface is also not clearly understood. When a person thinks to himself "I will turn on the light," his hand moves to the light switch and moves the switch in the appropriate direction. If we don't understand how thoughts become electrical impulses in the nerves, it is difficult to design an electromechanical device that can respond properly to such thoughts.

Even though these complex electrical patterns are not completely understood, computers can interpret some of these signals. By using some of the simple electrical impulses that are under conscious control, scientists have wired test subjects to a music synthesizer hooked up to a computer. With conscious control and assistance from the computer, a subject can make adjustments to the sounds made by the synthesizer.

The brain still holds many mysteries that science is trying to unravel. Until brain wave patterns are better understood, the ability to directly use thought to control objects is a subject best left to novelists. Once the problems associated with connecting nerves and computers are worked out, scientists can then construct neural networks that combine human beings and computers.

One of the next frontiers of bionic research involves the use of porous silicon. Porous silicon is identical to the silicon used to make computer chips, except that it contains pores or openings. The addition of nerve cells to porous silicon, combined with light-sensitive photoelectric cells, could someday be used to make an artificial light-sensitive retina. Soon scientists will be able to connect nerves together through porous silicon chips. This will be the first step in creating a neural network.

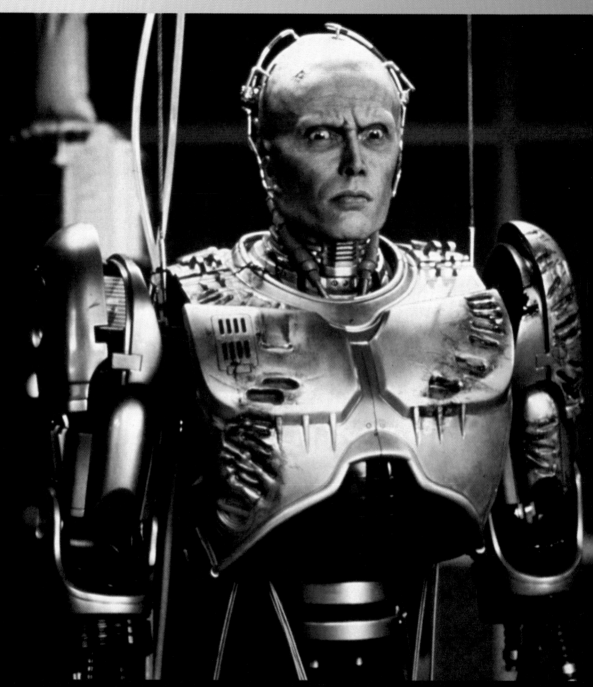

Cyborgs and androids—part human, part robot beings— have long been a popular element of science fiction movies and novels. One day they may become science fact.

6 The Future of Bionics

The term "cyborg" is short for cybernetic organism. A cyborg is usually defined as an individual with a high percentage of computerized electromechanical body parts. Cyborgs are a common presence in science fiction. In stories, cyborgs usually have greater powers than ordinary humans. In reality, a person with an artificial arm or leg could be called a cyborg, and even with the best of prosthetic appliances such people have limited abilities compared to people who have all their limbs intact. Someone who wears glasses could be considered a cyborg. It's all a matter of definition.

Science fiction writers, drawing not only on the fields of bionics and organ replacement but on new research in genetic engineering, have explored the wildest possibilities. They have imagined not simply human beings with restored limb function or improved sensory abilities, but human beings "customized" to survive in completely different environments. They have imagined human beings with gills to breathe under water and genetically modified people who could live on Mars. How about people with radiation-resistant tissues to work in nuclear laboratories and power stations? All this, of course, is so far beyond current medical technology that we might as well call it magic. And it raises complex ethical issues about the modification of the human organism and what it means to be human.

The ability to install working bionic body parts is now considered a routine procedure in many cases. The ability to completely rebuild a human being with a high percentage of artificial limbs, sensory apparatus, and internal organs is rather unlikely. Certain organs, like the heart, will continue to receive attention because developments in this area preserve and extend

Professor Sundro Mussa-Lvaldi holds the tiny cyborg device he invented in 2001. The mechanical part of the robot is controlled by the brain of a lamprey eel.

human life. But will the effort focus on the development of a permanent pump to replace the heart completely, or simply a more reliable pump used as a temporary support until a real heart becomes available? This depends on how rapidly other areas of medicine develop. There is some hope that new heart muscle tissue can be grown artificially and then implanted to replace diseased heart tissue. If procedures like this became commonplace, it might become unnecessary to develop an artificial heart designed to work on a permanent basis.

As long as human beings live in a world where accidents can occur, there will be a need for bionic arms and legs, and advanced designs that provide more mobility or sensitivity will always be in demand. Problems with sight and hearing, however, may be increasingly handled by advanced surgical techniques or specific cures for some of the diseases that afflict the eyes and the ears. In spite of the marvelous promise of technology, it is difficult to conceive of artificially constructed devices that can work as well as these organs.

Until there is an increase in the number of organs available for transplant, researchers will continue to work on artificial replacement organs for many of the organs that commonly fail us as we age or get sick. The technology is likely to be very expensive, and both the pubic and the medical profession will have to confront the thorny problem of how much should be spent to prolong human life.

Glossary

biomaterial Any material that is compatible with biological tissues and does not damage those tissues or cause an adverse reaction from the immune system.

bionic Used to describe the study of anatomy and physiology for the purpose of adapting or imitating biological functions with electronic or electro-mechanical designs.

congenital birth defect An abnormality that is present at birth.

cyborg An organism made up of both human and mechanical parts.

electric potential A difference in electromotive force, which creates a voltage and forces electric current to flow.

hybrid organ An organ that is made up of living cells and biomaterials.

myoelectric Electric potentials that are present in the muscles as they contract or relax.

neural network A series of nerves that are connected together.

polymer A long chain of short, repeating molecular units that are bonded together.

prosthesis An artificial device that replaces a missing biological part.

regeneration The ability to regrow.

For More Information

ORGANIZATIONS

The American Society for Artificial Internal Organs
P.O. Box C
Boca Raton, FL 33429-0468
(561) 391-8589
Web site: http://www.asaio.com

The Amputee Coalition of America
900 East Hill Avenue, Suite 285
Knoxville, TN 37915-2568
(888) AMP-KNOW (267-5669)
Web site: http://www.amputee-coalition.org

Institute for Cognitive Prosthetics
P.O. Box 892
Bala Cynwyd, PA 19004
(800) 837-5640
e-mail: icpinfo@brain-rehab.com
Web site: http://www.brain-rehab.com

Prosthetics Research Laboratory and Rehabilitation
    Engineering Research Program—Northwestern
    University
345 East Superior Street, Room 1441
Chicago, IL 60611-4496
e-mail: reiu@northwestern.edu
Web site: http://www.repoc.northwestern.edu

The Society for Biomaterials
13355 Tenth Avenue North, Suite 108
Minneapolis, MN 55441-5510
(763) 543-0908
Web site: http://www.biomaterials.org

WEB SITES

Due to the changing nature of Internet links, the Rosen Publishing Group, Inc., has developed an online list of Web sites related to the subject of this book. This site is updated regularly. Please use this link to access the list:

http://www.rosenlinks.com/lfm/bihu/

For Further Reading

Darling, David J. *The Health Revolution: Surgery and Medicine in the Twenty-first Century.* Parsippany, NJ: Dillon Press, 1996.

Metos, Thomas H. *Artificial Humans: Transplants and Bionics.* New York: J. Messner, 1985.

Murphy, Wendy B. *Spare Parts: From Peg Legs to Gene Splices.* Brookfield, CT: Twenty-First Century Books, 2001.

Wilson, A. Bennet, Jr. *A Primer on Limb Prosthetics.* Springfield, IL: Charles Thomas, 1998.

Bibliography

Beecroft, Simon. *Superhumans: A Beginner's Guide to Bionics*. Brookfield, CT: Copper Beech Books, 1998.

Berger, Melvin. *Bionics*. New York: Watts, 1978.

Dario, Paolo, Giulio Sandini, and Patrick Aebischer. *Robots and Biological Systems: Towards a New Bionics?* New York: Springer-Verlag, 1993.

Halacy, Daniel Stephen. *Cyborg: Evolution of the Superhuman*. New York: Harper & Row, 1965.

Index

Credits

ABOUT THE AUTHOR

Allan B. Cobb is a freelance science editor who lives in central Texas. He has written books, articles, radio scripts, and educational materials concerning different aspects of science. When not writing about science, he enjoys traveling, camping, hiking, and exploring caves.

PHOTO CREDITS

Cover © Mike Derer/AP Wide World Photos; Cover inset (front and back), p. 1 © PhotoDisc, Getty Images; Folio banners © EyeWire; pp. 4–5, 16–17 © Roger Ressmeyer/Corbis; pp. 6–7 © Mehau Kulyk/Science Photo Library; p. 9 by Evelyn Horovicz; p. 11 © David Stephenson/*Lexington Herald-Leader*/AP Wide World Photos; p. 12 © Michael Freeman/Corbis; pp. 14, 22 © Ed Kashi/Corbis; p. 20 © Steven Senne/AP Wide World Photos; p. 24 © Marta Lavandier/AP Wide World Photos; pp. 26–27 © Gary Emord-Netzley/*Messenger-Inquirer*/AP Wide World Photos; p. 31 © Stephen Charnin/AP Wide World Photos; p. 35 ©Volker Steger/Peter Arnold, Inc.; pp. 36–37 © John Lair/Jewish

DESIGN AND LAYOUT

Evelyn Horovicz